FESTUNG GUERNSEY

This book is unique amongst all volumes on German fortifications of the Second World War, in that it was written by the German forces themselves. It is the ultimate guide to their defensive planning and mindset.

The Channel Islands of Guernsey and Jersey nestle in the lee of the French coast. They are, however, British and in 1940 they became the first and only British territory to be occupied by the rapidly moving armies of Hitler's blitzkrieg.

Hitler swiftly became obsessed by his conquest. Determined that they should not be retaken he set in place a massive series of fortifications designed to make the islands impregnable fortresses or Festung, as part of the Atlantic Wall.

The Atlantic Wall ran for over three thousand miles, but by 1944 the tiny Channel Islands had consumed nearly 10% of all the concrete destined for the wall. They also contained the largest single garrison of the German Army and more heavy guns than 600 miles of Normandy coastline.

Field Marshall Rommel, in charge of the Normandy defences, pleaded with the Führer to release men and materials from the islands to Normandy. Hitler in a fury forbade him to raise the matter again and senior Reich officers began to talk of Hitler's "inselwahn" his "island madness".

In 1944 Lieutenant Colonel Hubner was charged with making a record of the immense fortifications. His team was drawn from the Divisionskartenstelle, the Divisional Cartographic Section, with some fourteen non-commissioned officers working across: drawing, photography, cartography, calligraphy and printing. The result is a stunning and comprehensive picture of the fortifications and a complete guide to their workings.

Festung Guernsey consists of 22 chapters and was originally published as a limited edition of 135, two-volume sets, bound in leather. The original work, being made by hand was only printed on the right hand pages, this means we have been able to provide a full translation on the left hand page, while retaining the original layout.

This paperback version will consist of 10 separate volumes, each consisting of 1,2 or 3 chapters and replicates the page numbering of the original edition.

Volumes are being published every four months with the 10th being completed in May 2015 on the 70th anniversary of the liberation of the islands by Force 135.

By 1944 Guernsey was the most fortified place in the world. These immense fortifications were built using slave labour. Please pause for a moment before turning the page and remember the men of many nationalities upon whose privations, ill treatment and lives this Festung was built.

The Festung Guernsey part work edition, part by part.

PART 1

Covers the coastal fortifications along the east coast
from St Martins Point to St Sampsons.

PART 2

**Covers the coastal fortifications along the east and north coasts
from St Samspsons to Grande Havre.**

PART 3

Covers the coastal fortifications along the west coast
from Grande Havre to Fort Sausmarez.

PART 4

Covers the coastal fortifications along the west and south coasts
from Rocquaine to Corbiere.

PART 5

The final coastal section covers the fortifications along the south
and east coasts from Corbiere to Fort George.

PART 6

The English Garrison on Guernsey, a history of the island's fortifications.

PART 7

General Information about Guernsey, a history of the island,
its laws and customs.

PART 8

Tactical Review of the Fortified areas & Fortified Structures,
detailing the garrison, and the bunker and emplacement designs.

PART 9

Weapons deployed & Mirus Battery, list and photographs of all the weaponry
and a chapter on the immense Mirus battery.

PART 10

Deployment of Artillery & Anti-Aircraft Artillery, comprises details and maps
of every artillery and anti-aircraft battery.

FESTUNG GUERNSEY

CHAPTERS: 3.3, 3.4 & 3.5

ST SAMPSON
BORDEAUX HARBOUR

from Mont Crevelt to Fort Doyle

ST SAMPSON
BORDEAUX-HAFEN
von Krevelberg bis Nebelhorn

St Sampson – Bordeaux Harbour
from Mont Crevelt to Fort Doyle

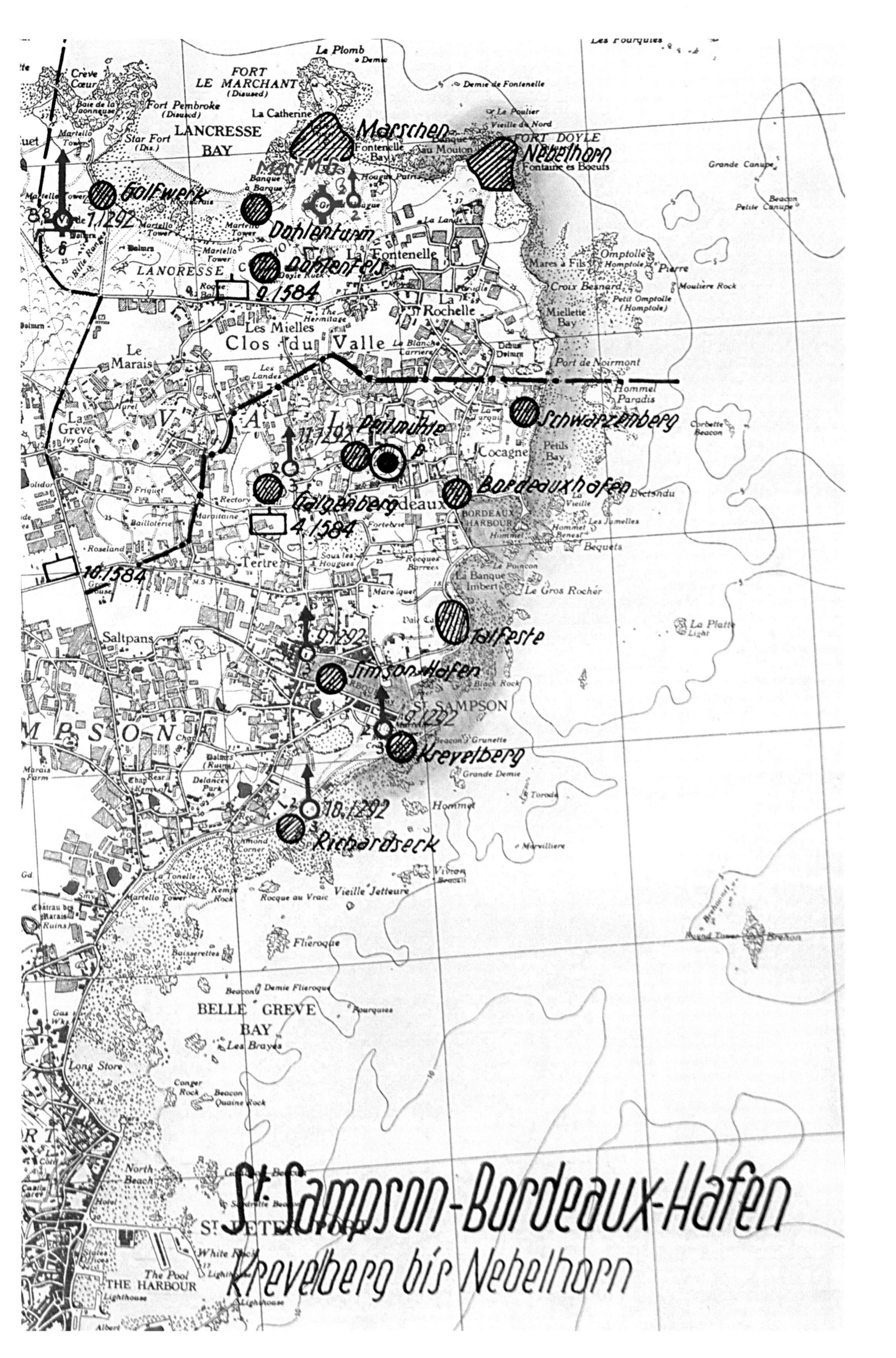
Marschen
Nebelhorn
Golfwerk
7./1292
Dohlenturm
Dohlenfels
9./1584
Schwarzenberg
11./1292
Pelmühle
Galgenberg
Bordeauxhafen
4./1584
10./1584
Talfeste
9./1292
Simson-Hafen
Krevelberg
18./1292
Richardseck
St. Sampson-Bordeaux-Hafen
Krevelberg bis Nebelhorn

Legend

- 10.5 cm gun in casemate
- 10.5 cm gun in field position
- 7.5 cm gun in embrasure
- 4.7 cm anti-tank gun in casemate
- 3.7 cm and 5 cm anti-tank gun in field position
- 3.7 cm tank gun
- 3.7 cm tank gun and machine-gun
- Armoured vehicle with machine-gun
- 8 cm and 5 cm mortars
- M19 (Maschinengranatwerfer, a fully-automatic 5 cm mortar)
- Tobruk pit
- Spigot mortar
- Multi-loopholed turret
- Searchlights
- Mines

Firing zones:

- 10.5 cm gun
- 4.7 cm anti-tank gun
- Anti-aircraft gun
- Mortar and M19
- Machine-gun
- Spigot mortar

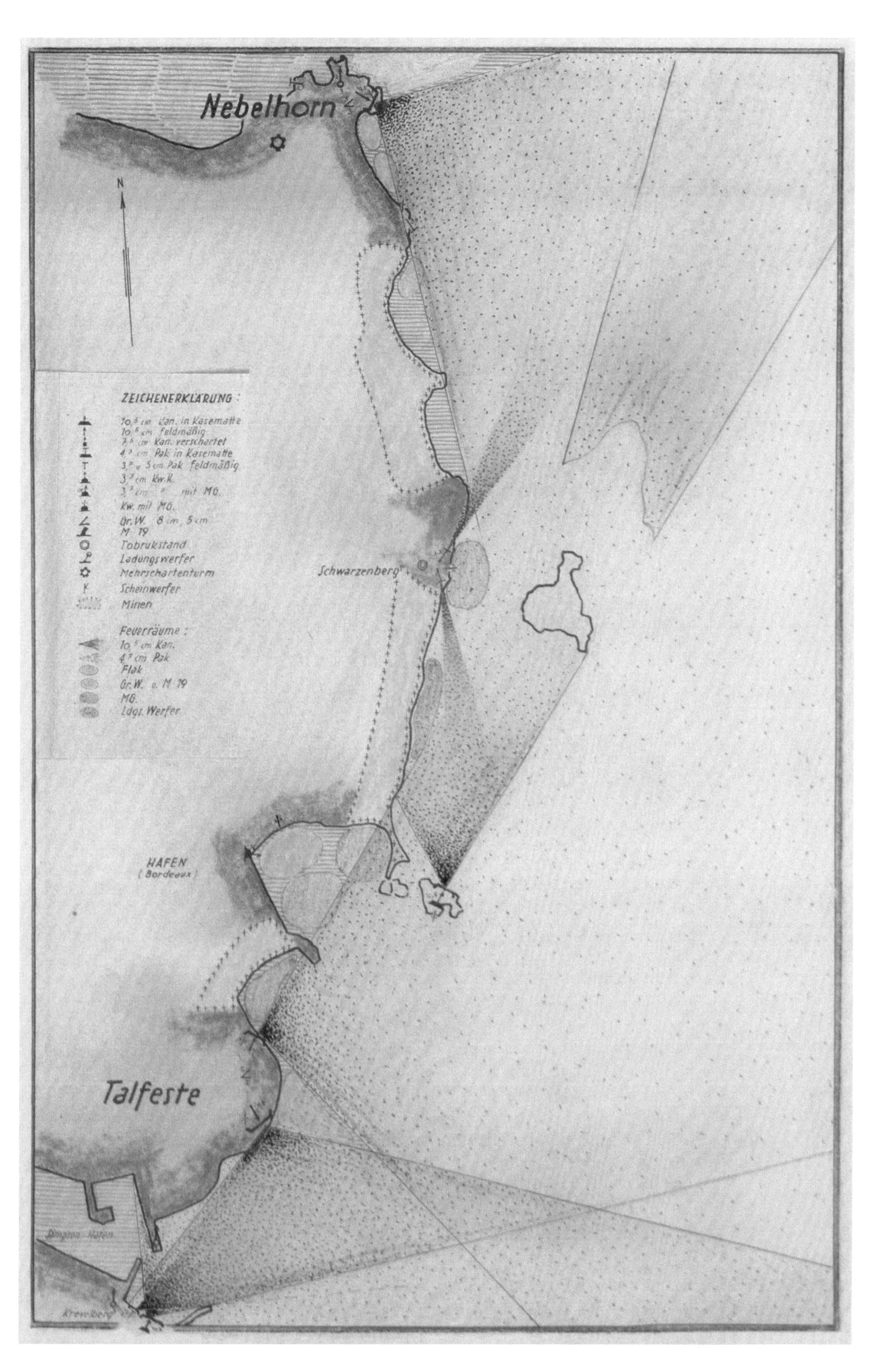
Nebelhorn
N
ZEICHENERKLÄRUNG :
10,5 cm Kan. in Kasematte
10,5 cm feldmäßig
7,5 cm Kan. verschartet
4,7 cm Pak in Kasematte
3,7 u. 5 cm Pak feldmäßig
3,7 cm Kw.K.
3,7 cm " mit MG.
Kw. mit MG.
Gr.W. 8 cm, 5 cm
M 19
Tobrukstand
Ladungswerfer
Mehrschartenturm
Scheinwerfer
Minen
Feuerräume :
10,5 cm Kan.
4,7 cm Pak
Flak
Gr.W. u. M 19
MG.
Ldgs. Werfer
Schwarzenberg
HAFEN
(Bordeaux)
Talfeste

ST SAMPSON HARBOUR Resistance Nest

1.) **Deployment** on Middle Mole at St Sampson Harbour between Mont Crevelt and Vale Castle.

2.) **Contingent**:

own:	1 NCO	7 men
navy:	1 petty officer	5 ratings
VGAD:	2 NCOs	11 men

3.) **Weapons**:

own:	1 machine gun MG311 (f) in tank turret 1 medium-range flame-thrower 3 defensive flame-throwers 42
navy:	2 2 cm anti-aircraft guns Z38 2 light machine guns MG34
VGAD:	1 light machine gun (Dutch) 1 light machine gun (b)

4.) **Military objectives**: The crew is charged with preventing enemy landings at St Sampson Harbour

5.) **Operations**: Enemy troops who have breached the harbour will be engaged using two light machine guns and the machine gun MG311 (f). The resistance nest will raise a counter-attack detachment to destroy enemy forces which have landed. Mechanised attacks will be repulsed by a tank destroyer squad.

In order to guard against low-level attacks by enemy aircraft, two 2 cm guns are to be deployed at the resistance nest.

In addition, an anti-tank mine-obstacle is located at the crossroads at the entrance to St Sampson Harbour, as protection against mechanised attack.

Widerstandsnest S I M S O N H A F E N

1.) Einsatz auf Mittelmole Hafen St.Sampson zwischen Krevelberg und Talfeste.

2.) Stärke:

eigene:	1 Uffz.	7 Mannschaften
Marine:	1 Uffz.	5 Mannschaften
VGAD:	2 Uffz.	11 Mannschaften

3.) Waffen:

eigene:	1	M.G.311 (f) in Panzerkuppel
	1	mittl. Flammenwerfer
	3	Abwehrflammenwerfer 42
Marine:	2	2 cm Flak Z 38
	2	le. M.G.34
VGAD:	1	le.M.G. (holl.)
	1	le.M.G. (b)

4.) Kampfauftrag: Die Besatzung hat den Auftrag, feindliche Landungen im Hafen St.Sampson zu verhindern.

5.) In den Hafen eingedrungener Gegner wird von 2 le.M.G. und 1 M.G.311 (f) bekämpft. Zur Vernichtung eines an Land gegangenen Gegners wird aus dem Widerstandsnest eine Gegenstoßgruppe herausgelöst. Angreifende Panzer werden durch einen Panzervernichtungstrupp bekämpft.

Zur Sicherung gegen Tieffliegerangriffe sind im Widerstands=nest 2 2 cm Geschütze eingesetzt.

Außerdem befindet sich an der Straßenkreuzung Eingang Sampson-Hafen zum Schutz gegen Panzerangriffe eine T-Minen-Schnellsperre.

VALE CASTLE Strong Point

1.) Deployment between St Sampson Harbour and Bordeaux Harbour.

2.) Contingent:

own:	2 NCOs	17 men
anti-aircraft artillery:		3 men
searchlight operation:		3 men

3.) Weapons:

own:	2 10.5 cm casemate guns (in field position) 2 light machine guns MG34 1 5 cm mortar 2 5 cm mortars (f) 1 medium-range flame-thrower 4 defensive flame-throwers 42
anti-aircraft:	1 60 cm searchlight
army:	1 60 cm searchlight

4.) Military objectives: The crew of the strong point will, in particular, repulse approaching enemy forces at St Sampson Harbour and Bordeaux Harbour. It will also fight off attempted enemy landings at Bordeaux Harbour.

5.) Operations: Enemy troops approaching the entrance to St Sampson Harbour will be engaged using a casemate gun and a heavy machine gun. Enemy craft attempting to put in at Bordeaux Harbour will be fired on using one of the 10.5 cm casemate guns.

Enemy forces landing at the bay between Vale Castle and Bordeaux Harbour will be engaged using a heavy machine gun and three 5 cm mortars. At the rear of the strong point there is meadowland intersected by hedges, suitable for dropping paratroopers. Three 5 cm mortars located at Vale Castle can be used to cover this area.

Tanks attacking from the land side will be engaged by an anti-tank squad.

A counter-attack detachment from the base will be made available to eliminate enemy forces who have established themselves in the area.

Stützpunkt T A L F E S T E

1.) Einsatz zwischen Simson- und Bordeaux-Hafen.

2.) Stärke:

eigene:	2 Uffz.	17 Mannschaften
Flak:		3 Mannschaften
Scheinwerf.Bedienung:		3 Mannschaften

3.) Waffen:

eigene:	2	Kas.Kan. 1o,5 cm (feldmäßig)
	2	le.M.G.34
	1	5 cm Gr.Werf.
	2	5 cm Gr.Werf. (f)
	1	mittl. Fla.-Werfer
	4	Abwehrflammenwerfer 42
Flak:	1	6o cm Scheinwerfer
Heer:	1	6o cm Scheinwerfer

4.) Kampfauftrag: Die Besatzung bekämpft insbesondere die Annäherung von Feindkräften vor St.Sampson-und Bordeaux-Hafen, sowie Anlandungen im Bordeaux-Hafen.

5.) Kampfführung: Sich der Hafeneinfahrt St.Sampson nähernde Feindkräfte werden von einer Kas.Kan. und einem s.M.G. bekämpft. Feindboote, die versuchen, den Bordeauxhafen anzulaufen, werden von einer Kas.Kan. 1o,5 cm unter Feuer genommen.

In der Bucht zwischen Talfeste und Bordeauxhafen an Land gehender Feind wird mit einem s.M.G. und 3 5 cm Gr.Werf. bekämpft. Im Hintergelände befindet sich ein zum Absprung von Fallschirmjägern geeignetes, von Hecken durchzogenes Wiesengelände. In diesen Abschnitt können von Talfeste 3 5 cm Gr.Werf. eingesetzt werden.

Von Landseite angreifende Panzer werden durch einen Panzernahkampftrupp bekämpft.

Zur Vernichtung eines Gegners, der sich im Gelände festgesetzt hat, wird eine Gegenstoßgruppe von dem Stützpunkt angesetzt.

BORDEAUX HARBOUR Resistance Nest

1.) **Deployment** at the deepest inlet at Bordeaux Harbour, below Peilmühle and between Vale Castle and Port de Noirmont.

2.) **Contingent**: 1 NCO 9 men

3.) **Weapons**: 1 3.7 cm gun and machine gun MG311 (f) in tank turret
1 light machine gun MG34

4.) **Military objectives**: The crew will repulse attempted enemy landings at Bordeaux Harbour.

5.) **Operations**: Enemy forces breaching Bordeaux Harbour will be engaged using a 3.7 cm gun and machine gun in tank turret, and a light machine gun in a Tobruk pit. Access to the harbour is impeded by a robust floating boom.

Beach obstructions at Bordeaux Harbour, consisting of iron girders and built-in 27 cm grenades, will prevent enemy craft entering the harbour at high tide. At low tide, the obstructions serve as obstacles against airborne troops and tanks.

The resistance nest is supported by the fire power of a spigot mortar, an 8 cm mortar and a heavy machine gun, all positioned at Peilmühle, and three 5 cm mortars stationed at Vale Castle.

A tank destroyer squad will repel mechanised attacks.

Enemy forces who have penetrated the terrain at the rear will be eliminated by a counter-attack detachment.

Widerstandsnest B O R D E A U X H A F E N

1.) Einsatz an der tiefsten Einbuchtung des Bordeauxhafens unterhalb Peilmühle zwischen Talfeste und Schwarzenberg.

2.) Stärke: 1 Uffz. 9 Mannschaften

3.) Waffen: 1 3,7 cm Kw.K. mit M.G.311 (f) in Panzerkuppel
1 le.M.G.34

4.) Kampfauftrag: Die Besatzung bekämpft Anlandung von Feindkräften im Bordeauxhafen.

5.) Kampfführung: Ein in den Bordeauxhafen eingedrungener Gegner wird mit einer 3,7 cm Kw.K. mit M.G. in Panzer=kuppel und einem le.M.G. aus Tobrukstand bekämpft. Die Einfahrt in den Hafen wird durch eine starke Balken=sperre erschwert.

Eine im Bordeauxhafen eingebaute Vorstrandsperre aus Eisenträgern und eingebauten 27 cm Granaten verhindert das Eindringen feindlicher Boote bei Flut. Bei Ebbe wirkt sie als Luftlande- und Panzersperre.

Das Widerstandsnest wird in seinem Feuerkampf unter=stützt durch einen Ladungswerfer, einen 8 cm Gr.Werf. und ein s.M.G. von Peilmühle sowie drei 5 cm Gr.Werf. von Talfeste.

Angreifende Panzer werden durch einen Panzervernichtungs=trupp bekämpft.

Ins Hintergelände eingedrungener Feind wird durch eine Gegenstoßgruppe vernichtet.

PORT DE NOIRMONT Resistance Nest

1.) **Deployment** at the boundary between the left-hand company and the battalion at Schwarze Düne between Bordeaux Harbour and Fort Doyle.

2.) **Contingent**: 2 NCOs 8 men

3.) **Weapons**:

2 3.7 cm anti-tank guns
1 heavy machine gun MG34
1 machine gun MG311 (f) in tank turret
1 handheld searchlight

4.) **Military objectives**: The crew of the resistance nest will repulse enemy landings at Schwarze Düne and enemy troops which have landed at Hommet Paradis.

5.) **Operations**: Enemy forces approaching off the coastline between Bordeaux Harbour and Port de Noirmont will be fired on using a 3.7 cm anti-tank gun. The resistance nest's right-hand flank is protected by a minefield.

Hommet Paradis is located approximately 500 metres offshore from the resistance nest and is mined with 27 cm grenades. These will be detonated by remote control from the resistance nest, as required.

Enemy forces attacking from the direction of Hommet Paradis will be engaged using two 3.7 cm anti-tank guns and a heavy machine gun.

The resistance nest's left-hand flank, which is also the left boundary of company and battalion, is protected by a minefield located along that boundary.

The resistance nest is protected against mechanised attack from the rear by:

a) a mined obstacle located where the road forks 100 metres north-west of Port de Noirmont;
b) two 3.7 cm anti-tank guns in pre-prepared alternative emplacements;
c) a tank destroyer squad, which can be deployed at pre-prepared positions or flexibly.

In addition, a machine gun MG311 (f) in tank turret will cover the area to the rear of the resistance nest.

Widerstandsnest S C H W A R Z E N B E R G

1.) Einsatz an linker Kompanie und Btls.-Grenze auf Schwarzer Düne zwischen Bordeauxhafen und Nebelhorn.

2.) Stärke: 2 Uffz. 8 Mannschaften

3.) Waffen:
    2 3,7 cm Pak
    1 s.M.G.34
    1 M.G.311 (f) in Panzerkuppel
    1 Handscheinwerfer.

4.) Kampfauftrag: Die Besatzung bekämpft die Anlandung von Feindkräften an Schwarzer Düne sowie angelandeten Feind auf Paradiesinsel.

5.) Kampfführung: Gegner, der sich im Küstensaum zwischen Bordeauxhafen und Schwarzenberg nähert, wird von einer 3,7 cm Pak unter Feuer genommen. Die rechte Flanke des Widerstandsnestes ist durch ein Minenfeld gesichert.

Die dem Widerstandsnest etwa 500 m vorgelagerte Paradies=insel ist mit 27 cm Granaten verseucht. Diese werden bei Bedarf vom Widerstandsnest aus ferngezündet.

Ein aus Richtung Paradiesinsel angreifender Feind wird mit 2 3,7 cm Pak und 1 s.M.G. bekämpft.

Die linke Flanke des Widerstandsnestes, gleichzeitig linke Kompanie- und Btls.-Grenze,ist gesichert durch ein auf der Naht befindliches Minenfeld.

Gegen von rückwärts angreifende Panzer ist das Wider=standsnest gesichert:

a) durch eine an der Wegegabel 100 m nordwestlich Schwarzenberg befindliche Schnellminensperre,
b) durch zwei 3,7 cm Pak aus vorbereiteten Wechsel=stellungen,
c) durch einen Panzervernichtungstrupp, der in vorberei=tete Stellungen und beweglich eingesetzt werden kann.

Außerdem ist nach rückwärts ein M.G.311(f) in Panzer=kuppel eingesetzt.

GALGENBERG Resistance Nest

1.) **Deployment** on the high ground between Peilmühle and Höhenviereck.

2.) **Contingent**: 1 officer 7 NCOs 26 men

3.) **Weapons**:
2 light machine guns MG34
1 heavy machine gun MG34
1 8 cm mortar
1 5 cm mortar
1 medium-range flame-thrower

4.) **Military objectives**: The crew will defend itself against airborne enemy troops landing between Höhenviereck and Galgenberg. On the company commander's special order, the men will be deployed either as assault detachments or as support forces for individual resistance nests within the company sector.

5.) **Operations**: Enemy forces attacking from the west, as well as paratroopers dropped in the area below Galgenberg, will be engaged using a spigot mortar, the 8 cm mortar, the heavy machine gun and the two light machine guns in Tobruk pits.

In the event of a successful breakthrough of enemy forces into the company sector, parts of the crew will be deployed as assault detachments on a flexible basis. The following weapons will be made available to them: two light machine guns, one 8 cm mortar and one set of anti-tank weapons.

Widerstandsnest G A L G E N B E R G

1.) Einsatz im Höhengelände zwischen Peilmühle und Höhenviereck.

2.) Stärke: 1 Offz. 7 Uffz. 26 Mannschaften

3.) Waffen:
   2 le.M.G.34
   1 s.M.G.34
   1 8 cm Gr.Werf.
   1 5 cm Gr.Werf.
   1 mittl. Flammenwerfer

4.) Kampfauftrag: Die Besatzung verteidigt sich gegen einen Gegner, der aus der Luft zwischen Höhenviereck und Galgenberg gelandet ist. Auf besonderen Befehl des Komp.-Führers wird sie stoßtruppartig oder als Verstär= kung einzelner Widerstandsnester im Kompanie-Abschnitt eingesetzt.

5.) Kampfführung: Ein aus Westen angreifender Gegner, sowie im Raum unterhalb Galgenberg abgesprungene Fallschirm= jäger werden mit einem Ladungswerfer, einem 8 cm Gr.W., einem s.M.G. und zwei le.M.G. aus Tobrukständen bekämpft.

   Bei gelungenem Feindeinbruch im Kompanie-Abschnitt werden Teile der Besatzung als Stoßtrupp beweglich eingesetzt. Dazu stehen folgende Waffen zur Verfügung: 2 le.M.G., 1 8cm Gr.W. und 1 Satz Panzernahkampfmittel.

PEILMÜHLE Resistance Nest

1.) **Deployment** on the high ground to the west of Bordeaux Harbour.

2.) **Contingent**: 1 NCO 12 men

3.) **Weapons**:

1 20 cm spigot mortar
1 8 cm mortar
1 light machine gun MG34
1 medium-range flame-thrower
2 handheld searchlights

4.) **Military objectives**: The crew will repulse enemy forces attacking from the direction of Port de Noirmont and Bordeaux Harbour, as well as preventing the landing of paratroopers on the meadowland 300 metres to the south of the resistance nest.

5.) **Operations**: Enemy forces attacking from the direction of Bordeaux Harbour and Port de Noirmont will be engaged using the spigot mortar, the 8 cm mortar and a heavy machine gun. Paratroopers who have been dropped between Vale Castle, Bordeaux Harbour and Peilmühle are also to be engaged using above-mentioned weapons.

Defence of the resistance nest will be supported by: three 5 cm mortars located at Vale Castle and one 8 cm mortar at Galgenberg.

A tank destroyer squad will repel mechanised attacks.

Enemy forces which have established themselves in the area will be eliminated by a counter-attack detachment from the resistance nest.

Widerstandsnest P E I L M Ü H L E

1.) Einsatz im Höhengelände westl. Bordeauxhafen.

2.) Stärke: 1 Uffz. 12 Mannschaften

3.) Waffen:
1 Ladungswerfer 2o cm
1 8 cm Granatwerfer
1 le.M.G.34
1 mittl. Flammenwerfer
2 Handscheinwerfer

4.) Kampfauftrag: Die Besatzung bekämpft aus Richtung Schwarzenberg und Bordeauxhafen angreifenden Feind, sowie Absprung feindlicher Fallschirmjäger im Wiesen=gelände 300 m südlich des Widerstandsnestes.

5.) Kampfführung: Angreifender Feind aus Richtung Bordeaux-hafen und Schwarzenberg wird mit einem Ladungswerfer, einen 8 cm Granatwerfer und ein s.M.G. bekämpft. Zwischen Talfeste, Bordeauxhafen und Peilmühle abgesprungene Fallschirmjäger werden ebenfalls mit oben genannten Waffen bekämpft.

Das Widerstandsnest wird in seinem Abwehrkampf unter=stützt von: 3 5 cm Gr.Werf. von Talfeste und 1 8 cm Gr.Werf. von Galgenberg.

Gegen angreifende Panzer wird ein Panzervernichtungs=trupp angesetzt.

Sich im Gelände festgesetzter Gegner wird durch eine Gegenstoßgruppe des Widerstandsnestes vernichtet.

FORT DOYLE Strong Point

1.) The Fort Doyle strong point has been established in the sector between Port de Noirmont and the middle of Fontenelle Bay.

2.) **Contingent of the base**:

Contingent of the base:	1 officer	4 NCOs	38 men
divided into:			
own:	1 officer	4 NCOs	34 men
naval anti-aircraft artillery:			3 ratings
searchlight operation:			1 man

3.) **Weapons**:

own weapons:	1 10.5 cm casemate gun (f)
	2 3.7 cm anti-tank guns
	1 5 cm mortar 36
	2 5 cm mortars 210 (f)
	1 heavy machine gun MG34
	1 machine gun on gun-carriage 08
	1 multi-loopholed armoured turret
	1 machine gun MG311 (f) in tank turret
assigned weapons:	3 2 cm anti-aircraft guns
	1 60 cm searchlight

4.) **Military objectives**: The strong point is charged with preventing enemy landings at Miellette Bay, Doyle Bay and Fontenelle Bay; with repulsing airborne troops landing at the base as well as in the area to the west of the base; and with repelling any attack from the land side.

5.) **Operations**: Enemy forces landing at the right flank will be engaged using the casemate gun and a 3.7 cm anti-tank gun, as well as the overlapping fire from two machine guns. Enemy troops attempting to land at Fontenelle Bay will be repulsed by a 3.7 cm anti-tank gun, a machine gun in tank turret and the turret with multiple embrasures. Dead spaces will be covered by the three light mortars. The Fort Le Marchant strong point will provide supporting fire at Fontenelle Bay with a casemate gun, a 5 cm anti-tank gun, two machine guns, an 8 cm mortar and two 5 cm mortars. Three 2 cm anti-aircraft guns will be used to guard against low-level attacks by enemy aircraft and airborne troops.

To repel mechanised attack from the rear, an augmented tank destroyer squad will be assigned. In addition, a 3.7 cm anti-tank gun will be transferred to a designated alternative emplacement.

Enemy troops landing outside the base will be engaged immediately by an assault detachment.

Stützpunkt N E B E L H O R N

1.) Der Stützpunkt Nebelhorn ist eingesetzt im Abschnitt von Schwarzenberg bis Mitte Forellenbucht.

2.) Stärke des Stützpunktes: 1 Offz. 4 Uffz. 38 Mannschaften
aufgeteilt:

eigene Kräfte:	1 Offz.	4 Uffz.	34 Mannschaften
Marine Flak:			3 Mannschaften
Scheinwerfer:			1 Mann

3.) Waffen:

eigene Waffen:
- 1 Kas.Kan. 1o,5 cm (f)
- 2 Pak 3,7 cm
- 1 Granatwerfer 36 (5 cm)
- 2 Granatwerfer 21o (f) (5 cm)
- 1 s.M.G.34
- 1 M.G. auf Lafette 08
- 1 Mehrschartenturm
- 1 M.G.311 (f) in Panzerkuppel

zugeteilte Waffen:
- 3 Flak 2 cm
- 1 Scheinwerfer 60 cm

4.) Kampfauftrag: Der Stützpunkt hat den Auftrag, jeden Lan=dungsversuch in der Mielette-, Doyle- und Forellenbucht (Fontenelle Bay) zu verhindern, Luftlandungen innerhalb und im Gelände westlich des Stützpunktes zu bekämpfen, sowie jeden Angriff von Land her abzuwehren.

5.) Kampfführung: In der rechten Flanke landender Gegner wird durch eine Kas.Kan., eine Pak 3,7 cm, sowie das sich überschneidende Feuer von 2 M.G., eine in der Forellenbucht versuchte Landung durch eine Pak 3,7 cm, ein M.G. in Panzer=kuppel und den Mehrschartenturm bekämpft. Tote Räume werden durch die drei leichten Granatwerfer ausgeschaltet. Der Stützpunkt Marschen unterstützt mit einer Kas.Kan., eine 5 cm Pak, zwei M.G., einen Granatwerfer 8 cm und zwei Granatwerfer 5 cm das Feuer in der Forellenbucht.
Zum Schutz gegen Tieffliegerangriffe und Luftlandungen sind 3 Flak 2 cm eingesetzt.
Zur Panzerabwehr nach rückwärts ist ein verstärkter Panzer=vernichtungstrupp eingeteilt. Außerdem bezieht eine Pak 3,7 cm eine hierfür vorbereitete Wechselstellung.
Gelandeter Gegner außerhalb des Stützpunktes wird sofort durch eine Stoßgruppe bekämpft.

Mont Crevelt strong point and St Sampson Harbour

STÜTZPUNKT KREWELBERG UND SAMPSON HAFEN

Top: Tug with minesweeper outside St Sampson Harbour
Middle: St Sampson Harbour – Herm in the background
Bottom: View from Vale Castle to St Sampson and Mont Crevelt at low tide

SCHLEPPER MIT M.-BOOT VOR SAMPSON HAFEN

SAMPSON-HAFEN. IM HINTERGRUND HERM

BLICK VON DER TALFESTE AUF SAMPSON U. KREWELBERG BEI NIEDRIGWASSER

St Sampson Harbour

SAMPSON-HAFEN

Top: Mont Crevelt strong point
Left Middle: View from Mont Crevelt to St Sampson Harbour and the town
Right Middle: Beach obstructions at St Sampson at low tide
Bottom: St Sampson Harbour with Vale Castle in the background

STÜTZPUNKT KREWELBERG

VOM KREWELBERG AUF SAMPSON-HAFEN U. STADT

VORSTRANDHINDERNISSE VOR SAMPSON BEI EBBE

SAMPSON HAFEN MIT DER TALFESTE IM HINTERGRUND

Top: Vale Castle
Left Middle: Entrance to Vale Castle. Hommet Paradis in the background
Right Middle: View from Vale Castle to Rungeturm and Jethou Island
Bottom: View from the mole entrance to St Sampson Harbour and the south-east corner of the island

DIE TALFESTE

EINGANG ZUR TALFESTE · IM HINTERGRUND D. PARADIESINSEL · VON DER TALFESTE ZUM RUNGETURM U. INSEL JETHOU

VOM EINGANG AUF DIE MOLE ZUM HAFEN SAMPSON UND DER SÜDOSTECKE DER INSEL

1: Beach obstructions at Mont Crevelt. On the left, bunker containing the mine ignition panel

2: Mole at St Sampson Harbour and Vale Castle

3: Entrance to St Sampson Harbour

4: View from Mont Crevelt to Herm, Jethou and Sark

VORSTRANDHINDERNISSE VOR KREVELBERG · LINKS BUNKER ZUR MINENZÜNDUNG

MOLE ZUM HAFEN SAMPSON MIT DER TALFESTE

HAFENEINFAHRT SAMPSON

VON KREVELBERG AUF HERM, JETHOU UND SARK

Top: Peilmühle ranging station
Middle: Aerial view of the Peilmühle strong point

MESS-STELLE „PEILMÜHLE"

STÜTZPUNKT PEILMÜHLE VON OBEN

Panoramas as sketched from the Peilmühle ranging station

ANSICHT-SKIZZEN AUF MESS-STELLE PEILMÜHLE

Panoramas as sketched from the Peilmühle ranging station

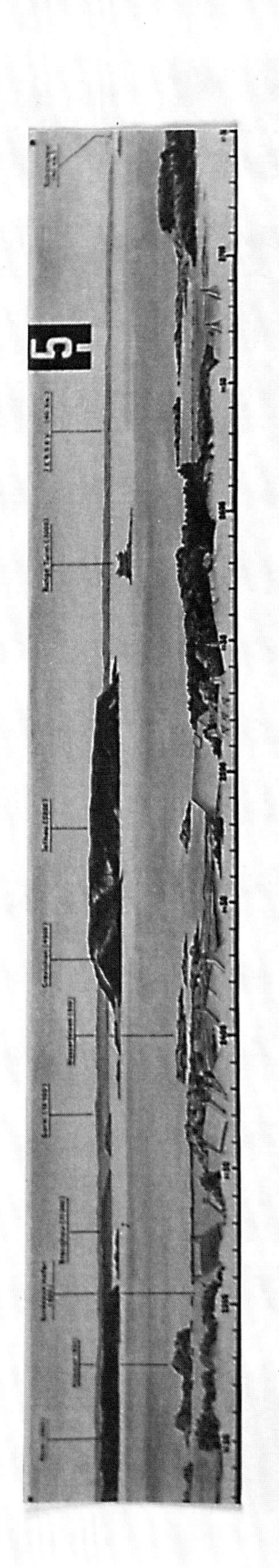

Panoramas as sketched from the Peilmühle ranging station

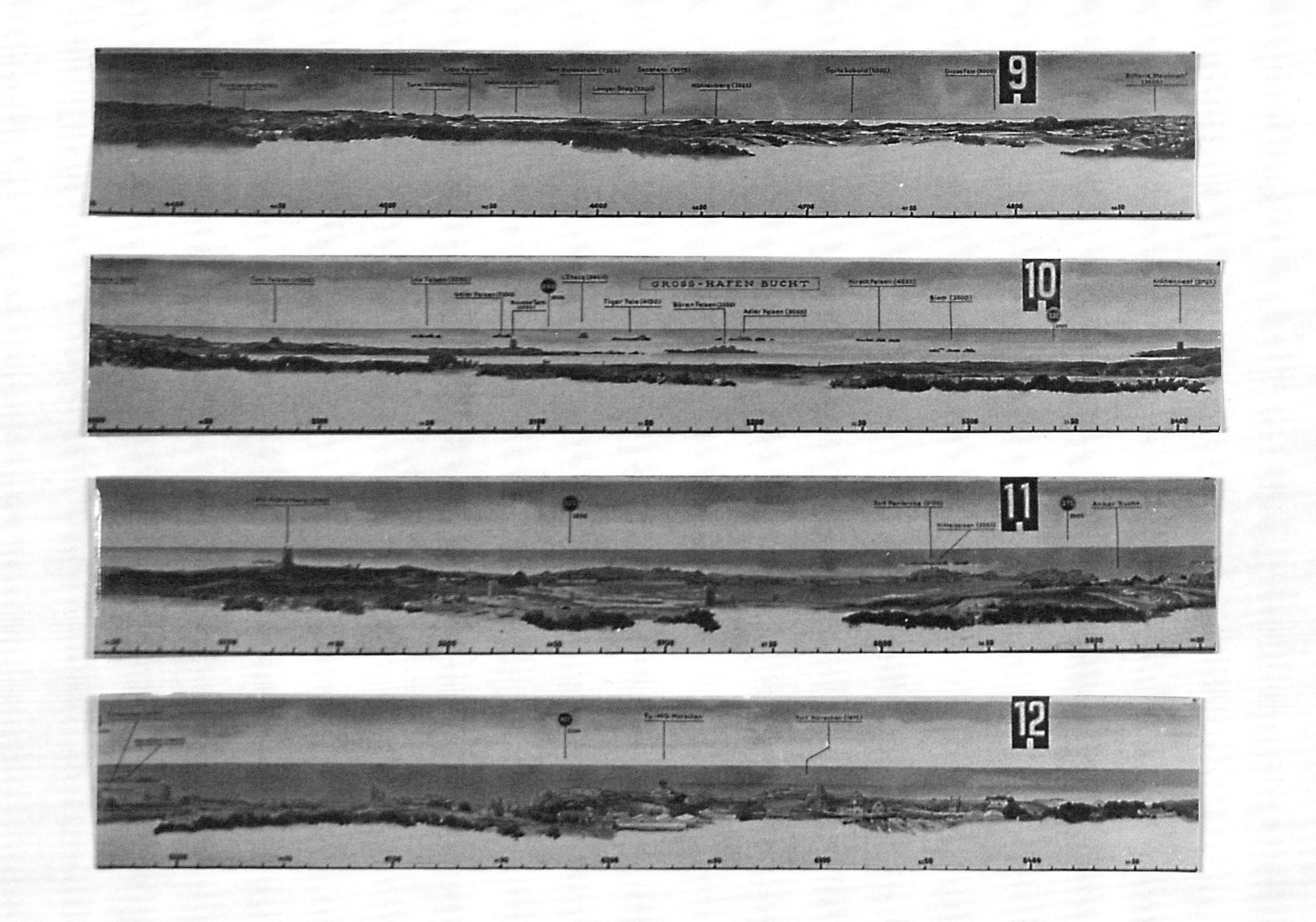
9
10
GROSS-HAFEN BUCHT
11
12

1: View from Vale Castle to Peilmühle
2: Entrance to Bordeaux Harbour at low tide
3: Fort Doyle strong point with bunker
4: Multi-loopholed armoured turret at Fort Doyle

BLICK VON DER TALFESTE ZUR PEILMÜHLE

EINFAHRT ZUM HAFEN BORDEAUX BEI EBBE

STÜTZPUNKT NEBELHORN MIT K.K. BUNKER

MEHRSCHARTENTURM AUF NEBELHORN

SAMPSON HAFEN UND KREVELBERG. IM HINTERGRUND DIE PEILMÜHLE UND TALFESTE

VON DER PEILMÜHLE AUF BORDEAUX, SAMPSON, TALFESTE, SCHÖNBUCHT UND DIE INSELN HERM, JETHOU, SARK

Top: St Sampson Harbour and Mont Crevelt; Peilmühle and Vale Castle in the background
Bottom: View from Peilmühle towards Bordeaux Harbour, St Sampson, Vale Castle, Belle Grève Bay and the islands of Herm, Jethou, Sark

Mont Crevelt, with entrance to St Sampson Harbour

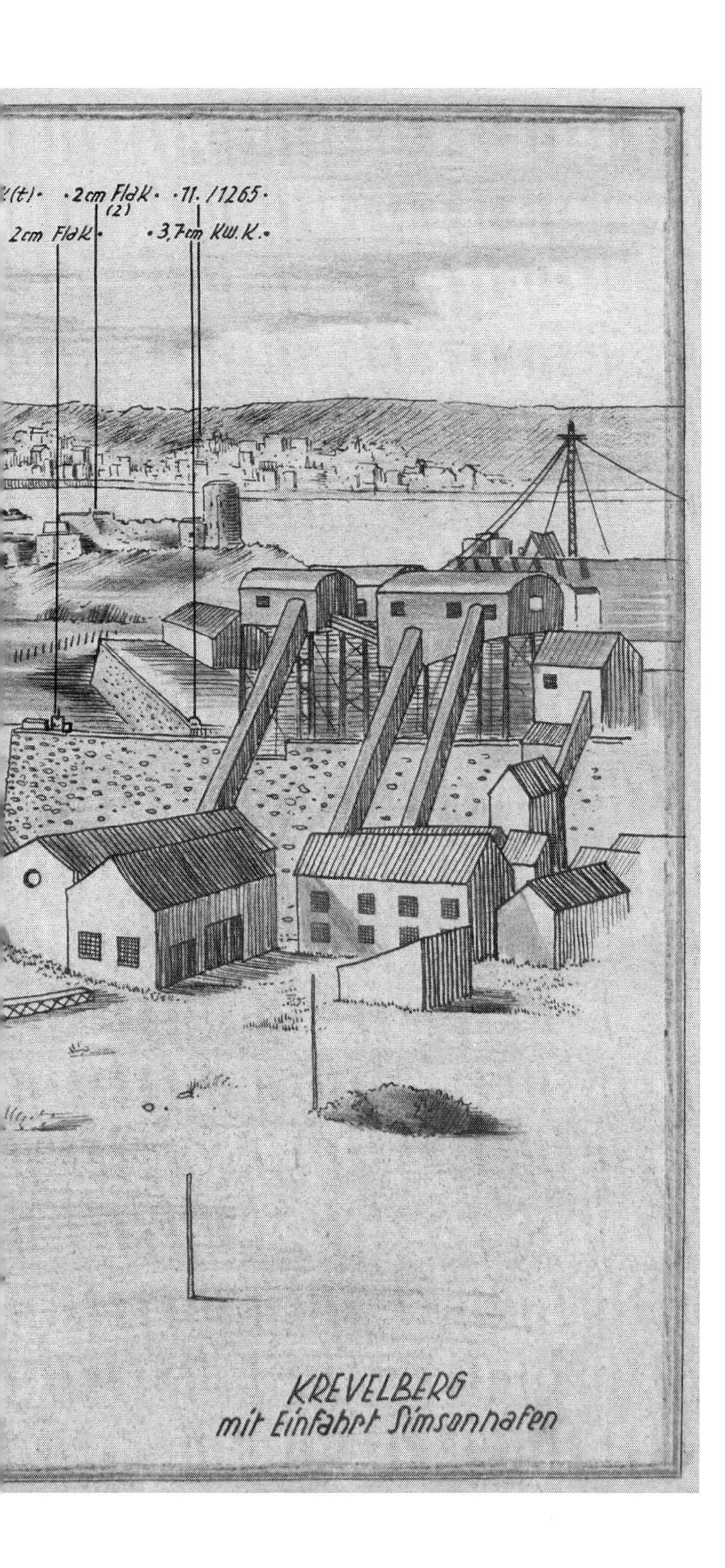

·2cm Flak·
(2)
·11. /1265·
2cm Flak·
·3,7cm KW.K.·
KREVELBERG
mit Einfahrt Simsonhafen

Vale Castle, with entrance to St Sampson Harbour

TALFESTE
mit Einfahrt Simsonhafen

L'ANCRESSE BAY

from Fort Doyle to La Varde

ANKERBUCHT

von Nebelhorn bis Golfwerk

L'Ancresse Bay
from Fort Doyle to La Varde

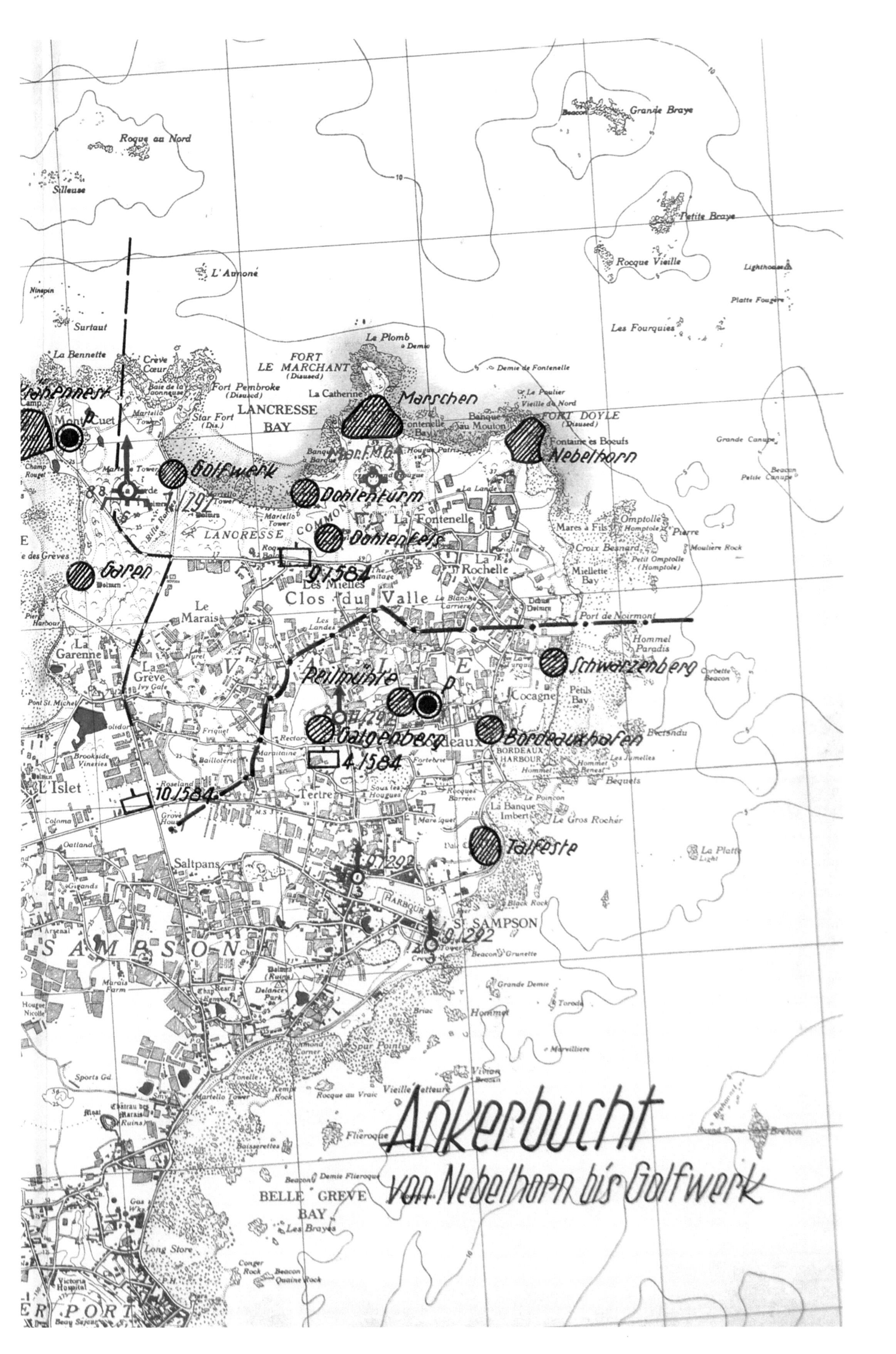

Marschen
Nebelhorn
Golfwerk
Dohlenturm
Dohlenfels
Garen
Schwarzenberg
Peilmühle
Galgenberg
Bordeauxhafen
Talfeste
9/584
4/584
10/584
7/292
9/292
8/292
Ankerbucht
von Nebelhorn bis Golfwerk

Legend

- 10.5 cm gun in casemate
- 10.5 cm gun in field position
- 7.5 cm gun in embrasure
- 4.7 cm anti-tank gun in casemate
- 3.7 cm and 5 cm anti-tank gun in field position
- 3.7 cm tank gun
- 3.7 cm tank gun and machine-gun
- Armoured vehicle with machine-gun
- 8 cm and 5 cm mortars
- M19 (Maschinengranatwerfer, a fully-automatic 5 cm mortar)
- Tobruk pit
- Spigot mortar
- Multi-loopholed turret
- Searchlights
- Mines

Firing zones:

- 10.5 cm gun
- 4.7 cm anti-tank gun
- Anti-aircraft gun
- Mortar and M19
- Machine-gun
- Spigot mortar

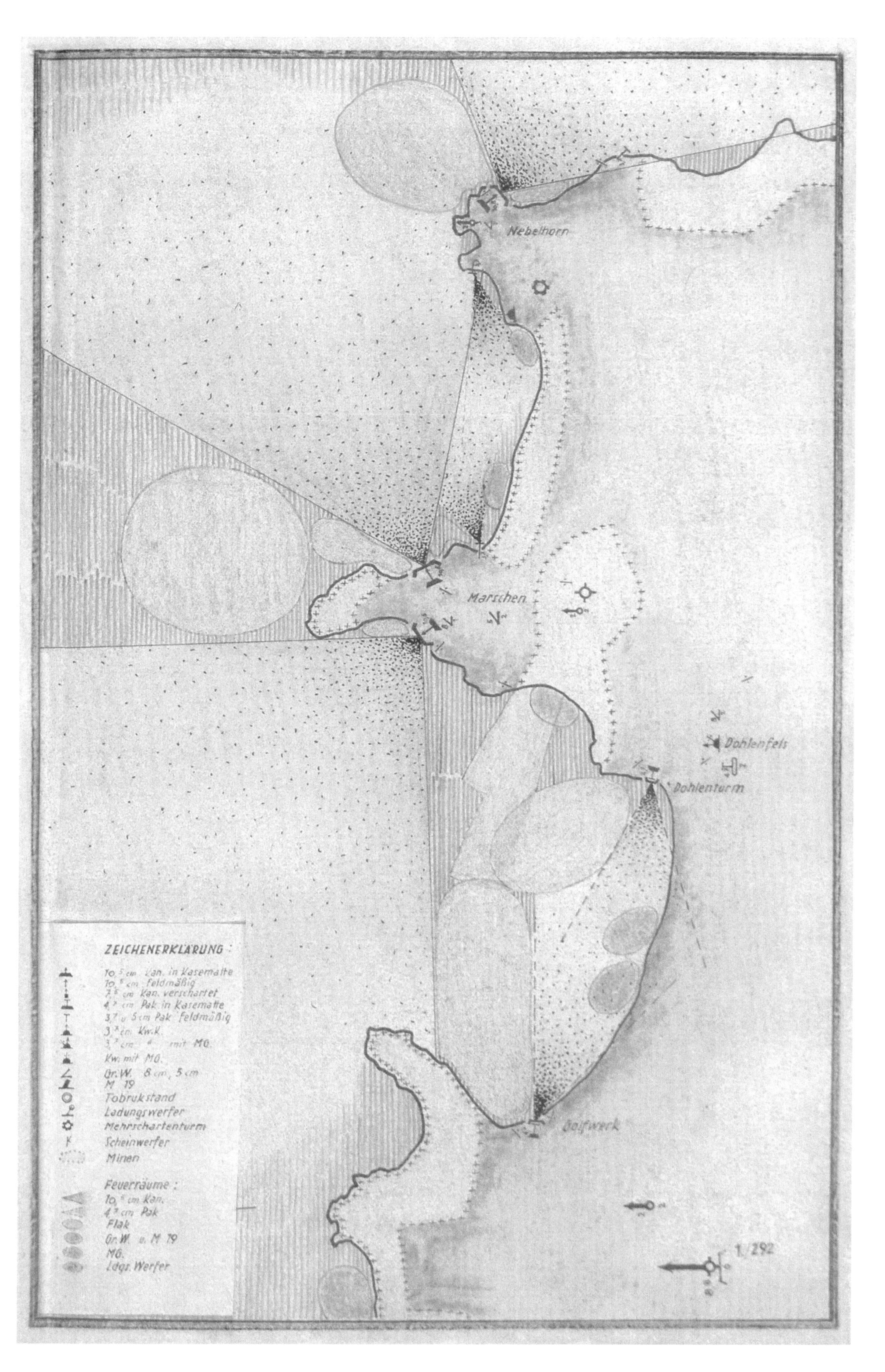
Nebelhorn
Marschen
Dohlenfels
Dohlenturm
Golfwerk
ZEICHENERKLÄRUNG :
10,5 cm Kan. in Kasematte
10,5 cm feldmäßig
7,5 cm Kan. verschartet
4,7 cm Pak in Kasematte
3,7 u. 5 cm Pak feldmäßig
3,7 cm Kw.K.
3,7 cm " mit MG.
Kw. mit MG.
Gr.W. 8 cm, 5 cm
M 19
Tobrukstand
Ladungswerfer
Mehrschartenturm
Scheinwerfer
Minen
Feuerräume :
10,5 cm Kan.
4,7 cm Pak
Flak
Gr.W. u. M 19
MG.
Ldgs. Werfer
1/292

FORT LE MARCHANT Strong Point

1.) The Fort Le Marchant strong point has been established in the sector between the middle of Fontenelle Bay and Banque à Barque in L'Ancresse Bay.

2.) **Contingent at the strong point**:

own forces:	5 NCOs	35 men
assigned forces:		
searchlight operation	1 NCO	2 men
radio operation		2 men

3.) **Weapons**:

own:	2 10.5 cm casemate guns (f)
	1 5 cm anti-tank gun
	1 8 cm mortar 34
	2 5 cm mortars 210 (f)
	1 heavy machine gun MG34
	3 machine guns MG34 on gun carriage 08
assigned:	2 60 cm searchlights

4.) **Military objectives**: The strong point is charged with preventing enemy landings at Fontenelle Bay and L'Ancresse Bay; to engage airborne troops immediately and to repel any attack from the land side.

5.) **Operations**: With its heavy and light weapons, the strong point possesses considerable firepower. Enemy troops landing at Fontenelle Bay are to be fought off using a casemate gun and the 5 cm anti-tank gun, supported by the weapons located at the Fort Doyle strong point. The beach and the coast along the northern tip of the base are covered by a machine gun each.

Enemy landing craft sailing into L'Ancresse Bay will be engaged using a casemate gun and the heavy machine gun.

The small bay between Fort le Marchant and L'Ancresse Tower is covered by fire from a machine gun and the 8 cm mortar.

The terrain at the rear of the strong point is eminently suitable for the landing of airborne troops. Enemy troops who have landed there will be attacked immediately by the base's assault detachment.

Two machine guns MG34 on twin swivel-mounts are available to guard against low-level attacks by enemy aircraft and airborne troops. Two 2 cm anti-aircraft guns located at the La Grande Hougue naval strong point will support the base's air defence.

The 5 cm anti-tank gun and a tank destroyer squad will be used to repel mechanised attack to the rear.

Stützpunkt M A R S C H E N

1.) Der Stützpunkt Marschen ist eingesetzt im Abschnitt von Mitte Forellenbucht (Fontenelle Bay) bis Banque a Barque in der Ankerbucht.

2.) Stärke des Stützpunktes:

eigene Kräfte:	5 Uffz.	35 Mannschaften
zugeteilte Kräfte:		
Scheinwerfer	1 Uffz.	2 Mannschaften
Funker		2 Mannschaften

3.) Waffen:

eigene:	2	Kas.Kan. 1o,5 cm (f)
	1	Pak 5 cm
	1	Granatwerfer 34 (8 cm)
	2	Granatwerfer 21o (f) (5 cm)
	1	s.M.G.34
	3	M.G.34 auf Lafette 08
zugeteilt:	2	Scheinwerfer 60 cm

4.) Kampfauftrag: Der Stützpunkt hat den Auftrag feindliche Landungsversuche in der Forellen- (Fontenelle Bay) und in Ankerbucht zu verhindern, Luftlandungen sofort zu bekämpfen und Angriffe von der Landseite abzuschlagen.

5.) Kampfführung: Der Stützpunkt hat mit seinen schweren und leichten Waffen eine große Feuerkraft. Durch die Waffen des Stützpunktes Nebelhorn unterstützt, bekämpft er landenden Gegner in der Forellenbucht mit einer Kas.Kan. und einer 5 cm Pak. Das Feuer von je einem M.G. liegt auf dem Strand und entlang der Nordspitze des Stützpunktes.
In die Ankerbucht einlaufende feindliche Landungsboote wer= den mit einer Kas.Kan. und einem s.M.G. bekämpft.
In der kleinen Bucht zwischen Marschen und Dohlenturm liegt das Feuer von einem M.G. und einem 8 cm Granatwerfer.
Das Gelände rückwärts des Stützpunktes eignet sich gut für Luftlandungen. Durch die Stoßgruppe des Stützpunktes wird gelandeter Gegner sofort angegriffen.
Zur Sicherung gegen Tieffliegerangriffe und Luftlandungen sind 2 M.G.34 auf Zwillingssockel vorhanden. 2 Flak 2 cm des Marine-Stützpunktes Großhügel unterstützen die Luftab= wehr des Stützpunktes.
Zur Panzerabwehr nach rückwärts wird die 5 cm Pak und ein Panzervernichtungstrupp eingesetzt.

L'ANCRESSE TOWER Resistance Nest

1.) The L'Ancresse Tower resistance nest has been established in the sector between Banque à Barque and the middle of L'Ancresse Bay.

2.) **Contingent at the resistance nest**:

2 NCOs 7 men

3.) **Weapons**:

own: 1 4.7 cm anti-tank gun (t) with machine gun MG37 (t)
1 machine gun MG34 on gun carriage 08

4.) **Military objectives**: The resistance nest is charged with preventing enemy landings at L'Ancresse Bay.

5.) **Operations**: Given their flanking positions in L'Ancresse Bay, the resistance nests at L'Ancresse Tower and La Varde complement each other. Enemy landing craft and tanks and troops having disembarked at low tide will be engaged using the 4.7 cm anti-tank gun and a machine gun. The anti-tank wall stretching along the coast to the La Varde resistence nest constitutes an obstacle to enemy tanks.

The resistance nest at Doyle Rock will support this defence using two 8 cm mortars, a 3.7 cm gun in tank turret and a heavy machine gun.

An anti-tank squad equipped with anti-tank weapons is available to repel mechanised attack to the rear.

Widerstandsnest D O H L E N T U R M

1.) Das Widerstandsnest Dohlenturm ist eingesetzt im Abschnitt von Banque á Barque bis Mitte Ankerbucht.

2.) Stärke des Widerstandsnestes:
2 Uffz. 7 Mannschaften.

3.) Waffen:
eigene: 1 Pak 4,7 cm (t) mit M.G. 37 (t)
1 M.G.34 auf Lafette 08

4.) Kampfauftrag: Das Widerstandsnest hat den Auftrag feind= liche Landungen in der Ankerbucht zu verhindern.

5.) Kampfführung: Durch die flankierende Lage in der Anker= bucht ergänzen sich die Widerstandsnester Dohlenturm und Golfwerk. Feindliche Landungsboote und bei Ebbe bereits gelandete Panzer und Truppen werden von der 4,7 cm Pak und dem M.G. bekämpft. Die Panzermauer entlang der Küste bis zum Widerstandsnest Golfwerk bietet ein Hindernis für feindliche Panzer.

Das Widerstandsnest Dohlenfels unterstützt mit 2 8 cm Granatwerfern, einer Kw.Kan. 3,7 cm und einem s.M.G. den Feuerkampf.

Zur Panzerbekämpfung nach rückwärts ist ein Panzernah= kampftrupp mit Panzernahbekämpfungsmitteln vorhanden.

DOYLE ROCK Resistance Nest

1.) The Doyle Rock resistance nest has been established in the sector between La Fontenelle and the eastern edge of the Golf Course.

2.) **Contingent at the resistance nest**:

own:	3 NCOs	14 men

3.) **Weapons**:

own:

1 3.7 cm gun (f) in tank turret and machine gun MG 311 (f)
2 8 cm mortars 34
2 heavy machine guns MG34

4.) **Military objectives**: The resistance nest is charged with preventing enemy landings at L'Ancresse Bay and engaging airborne enemy troops.

5.) **Operations**: The resistance nest is to use its firepower to support the resistance nests at L'Ancresse Tower and La Varde. The two 8 cm mortars located at Doyle Rock will provide barrage fire covering the beach at L'Ancresse Bay and the junctions of road "Red 1". Enemy forces landed at L'Ancresse Bay or at the Golf Course will be engaged using the 3.7 cm gun (f) in tank turret and one of the heavy machine guns. The other heavy machine gun is allocated to defending the rear.

The Golf Course and the terrain between the L'Ancresse Tower and La Grande Hougue resistance nests are suitable for the landing of airborne troops. Enemy troops already landed will be engaged immediately by an assault detachment dispatched from this resistance nest, in conjunction with an assault detachment from the L'Ancresse Tower resistance nest.

An anti-tank squad equipped with anti-tank weapons is available to repel mechanised attack from the rear.

Widerstandsnest D O H L E N F E L S

1.) Das Widerstandsnest Dohlenfels ist eingesetzt im Abschnitt von La Fontanelle bis Ostrand Golfplatz.

2.) Stärke des Widerstandsnestes:

eigene: 3 Uffz. 14 Mannschaften

3.) Waffen:

eigene:
1 Kw.K. 3,7 cm (f) mit M.G.311(f) in Panzerkuppel
2 Granatwerfer 34 (8 cm)
2 s.M.G. 34

4.) Kampfauftrag: Das Widerstandsnest hat den Auftrag feind=liche Landung in der Ankerbucht zu verhindern sowie aus der Luft gelandeten Gegner zu bekämpfen.

5.) Kampfführung: Das Widerstandsnest unterstützt mit seinen Waffen die Widerstandsnester Dohlenturm und Golfwerk. Mit seinen zwei 8 cm Granatwerfern belegt es den Strand in der Ankerbucht und die Straßenkreuzungen der Straße Rot 1 mit Sperrfeuer. Mit der 3,7 cm Kw.K. und 1 s.M.G. bekämpft es gelandeten Gegner in der Ankerbucht und auf dem Golfplatz. Ein s.M.G. ist zur Rückwärtsverteidigung eingesetzt.

Der Golfplatz und das Gelände zwischen Widerstandsnest Dohlenturm und Großhügel eignet sich für Luftlandungen. Gelandeter Gegner wird sofort mit einer Stoßgruppe des Widerstandsnestes in Verbindung mit einer Stoßgruppe des Widerstandsnestes Dohlenturm angegriffen.

Zur Panzerbekämpfung nach rückwärts ist ein Panzernah=kampftrupp mit Panzernahbekämpfungsmitteln eingesetzt.

LA VARDE Resistance Nest

1.) The La Varde resistance nest has been established in the sector between the middle of L'Ancresse Bay and La Bennette.

2.) **Contingent at the resistance nest**:

assigned forces:	2 NCOs	8 men
(14th Battery, Rifle Regiment 584)		

3.) **Weapons**:

own: 1 4.7 anti-tank gun (t) with machine gun MG37 (t)
1 machine gun MG34 in Tobruk pit

4.) **Military objectives**: The resistance nest is charged with preventing enemy landings at L'Ancresse Bay.

5.) **Operations**: The resistance nests at La Varde and at L'Ancresse Tower complement each other in terms of their weaponry. Enemy forces already landed will be engaged using the 4.7 cm anti-tank gun and the machine gun. The anti-tank wall constitutes an obstacle to enemy tanks. The Doyle Rock resistance nest will provide supporting fire, using two 8 cm mortars, a 3.7 cm gun (f) in tank turret and a heavy machine gun.

Widerstandsnest G O L F W E R K

1.) Das Widerstandsnest Golfwerk ist eingesetzt im Abschnitt von Mitte Ankerbucht bis Benette-Bucht.

2.) Stärke des Widerstandsnestes:

zugeteilte Kräfte: 2 Uffz. 8 Mannschaften
(14./Gr.Rgt. 584)

3.) Waffen:

eigene: 1 Pak 4,7 cm (t) mit M.G. 37 (t)
1 M.G. 34 in Tobrukstand

3.) Kampfauftrag: Das Widerstandsnest hat den Auftrag feindlichen Landungsversuch in der Ankerbucht zu ver=hindern.

5.) Kampfführung: Das Widerstandsnest Golfwerk und Dohlenturm ergänzen sich gegenseitig mit ihren Waffen. Gelandeter Feind wird mit der 4,7 cm Pak und dem M.G. bekämpft. Die Panzermauer bietet für feindliche Panzer ein Hinder=nis. Das Widerstandsnest Dohlenfels unterstützt das Feuer mit 2 8 cm Granatwerfer, einer 3,7 cm Kw.Kan. und einem s.M.G.

Top: Commemorative plaque at Fort Le Marchant

Middle: View of L'Ancresse Bay from La Varde. The Fort Le Marchant strong point to the left, L'Ancresse Tower to the right.

Bottom: View of Fort Le Marchant as seen from the strong point

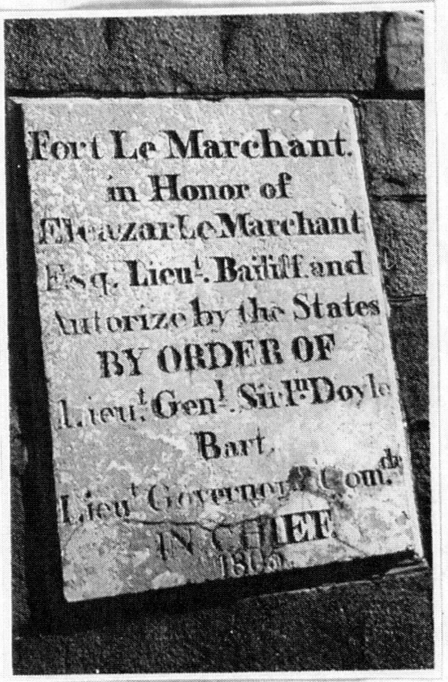

ERINNERUNGSTAFEL IM FORT MARCHAND

DIE ANKERBUCHT VOM GOLFWERK GESEHEN · LINKS STÜTZPUNKT MARSCHEN, RECHTS DER DOHLENTURM

FORT MARCHAND · GESEHEN VOM STÜTZPUNKT

Top: The entrance to Fort Le Marchant
Left Middle: Old bastion at the fort
Right Middle: Fort Le Marchant
Left Bottom: View of L'Ancresse Bay from the fort. L'Ancresse Tower to the left, La Varde to the right.
Right Bottom: Numerous old gun emplacements on the fort walls

EINGANG ZUM FORT MARCHAND

ALTE BASTION IM INNEREN FORT

FORT MARSCHEN (MARCHAND)

DIE ANKERBUCHT VOM FORT AUS · LINKS DER DOHLENTURM RECHTS GOLFWERK

VIELE ALTE GESCHÜTZSTÄNDE AUF DEN WÄLLEN

Top: Sentry at the entrance to the Fort Le Marchant strong point
Middle: The casemate gun: well camouflaged and protected

EINGANGSPOSTEN ZUM STÜTZPUNKT MARSCHEN

GUT GETARNT UND AUSGEBAUT DIE KAS.-KAN.

Top: At the Banque à Barque – La Garenne to the right
Middle: The anti-tank wall at L'Ancresse Bay. The ranging station at Mont Cuet in the background
Bottom: Obstructions on the Golf Course against airborne troops

AN DER BANQUE À BARQUE · RECHTS ORTSTEIL GAREN

DIE PANZERMAUER IN DER ANKERBUCHT
IM HINTERGRUND M.-STELLE KRÄHENBERG

LUFTLANDEHINDERNISSE AUF DEM GOLFPLATZ

L'Ancresse Bay

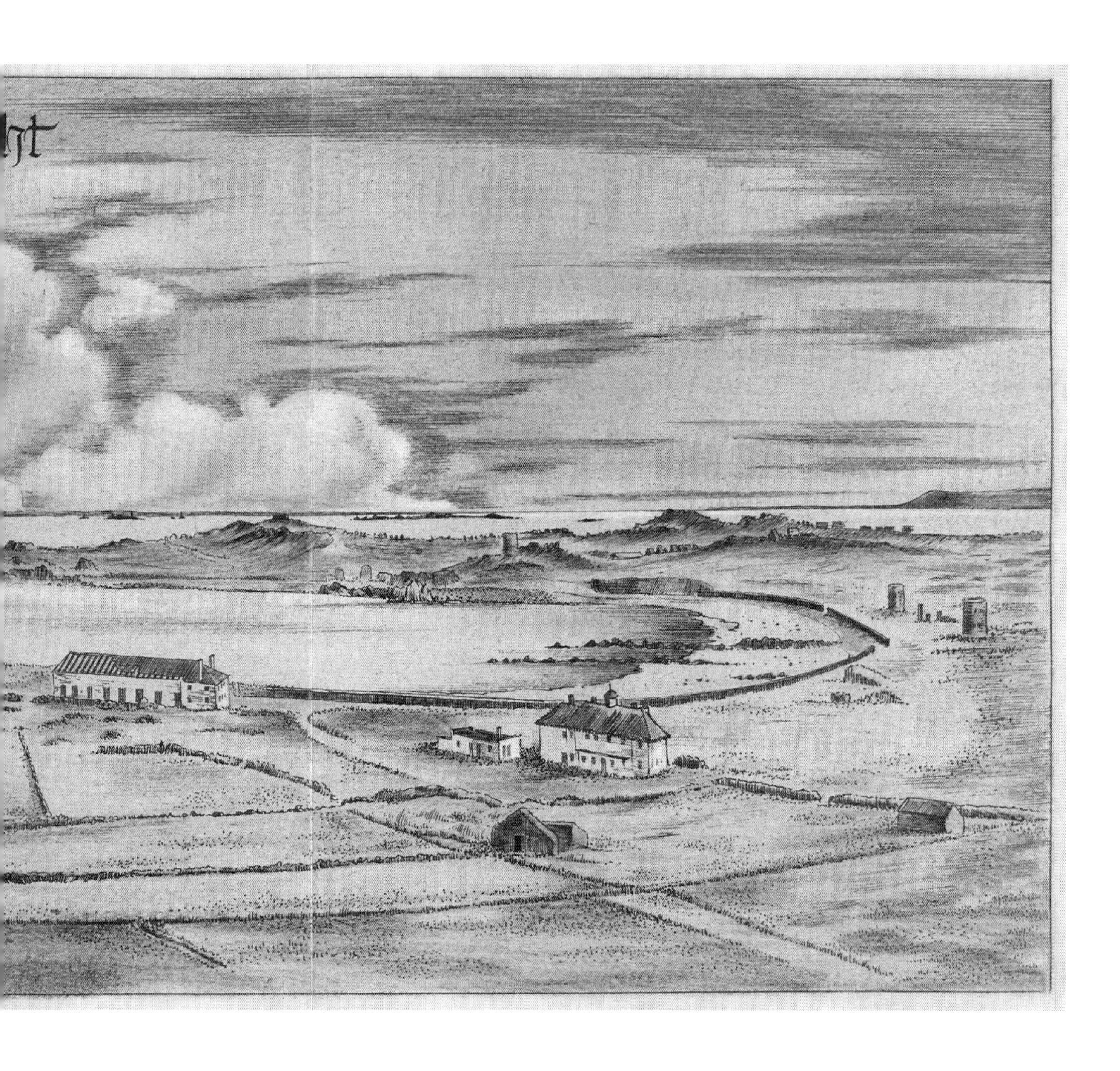

LE GRAND HAVRE

from Chouet to Le Houmet

GROSSBUCHT
von Krähennest bis Houmet

Le Grand Havre
from Chouet to Le Houmet

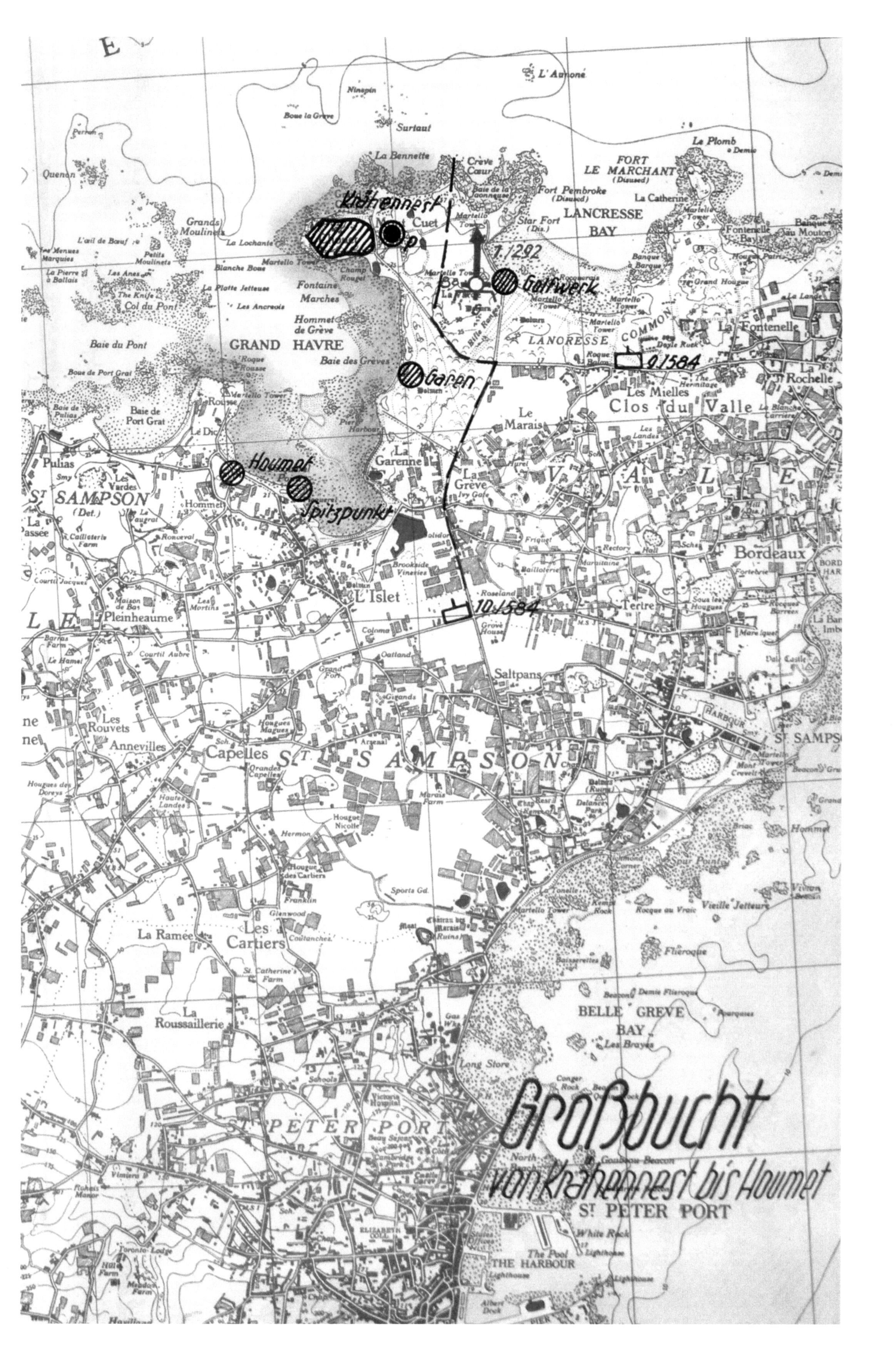

Großbucht
von Krähennest bis Houmet
Krähennest
Golfwerk
Garen
Houmet
Spitzpunkt
7.J292
9.J584
10.J584
FORT LE MARCHANT (Disused)
Fort Pembroke (Disused)
Star Fort
LANCRESSE BAY
GRAND HAVRE
Baie des Grèves
Baie du Pont
Baie de Port Grat
L'Ancresse Common
Les Mielles
Clos du Valle
VALE
ST SAMPSON
ST SAMPSON (Det.)
Pulias
L'Islet
Pleinheaume
Capelles
Les Cartiers
La Ramée
La Roussaillerie
Bordeaux
Saltpans
La Fontenelle
La Rochelle
BELLE GREVE BAY
PETER PORT
ST PETER PORT
THE HARBOUR

Legend

- 10.5 cm gun in casemate
- 10.5 cm gun in field position
- 7.5 cm gun in embrasure
- 4.7 cm anti-tank gun in casemate
- 3.7 cm and 5 cm anti-tank gun in field position
- 3.7 cm tank gun
- 3.7 cm tank gun and machine-gun
- Armoured vehicle with machine-gun
- 8 cm and 5 cm mortars
- M19 (Maschinengranatwerfer, a fully-automatic 5 cm mortar)
- Tobruk pit
- Spigot mortar
- Multi-loopholed turret
- Searchlights
- Mines

Firing zones:
- 10.5 cm gun
- 4.7 cm anti-tank gun
- Anti-aircraft gun
- Mortar and M19
- Machine-gun
- Spigot mortar

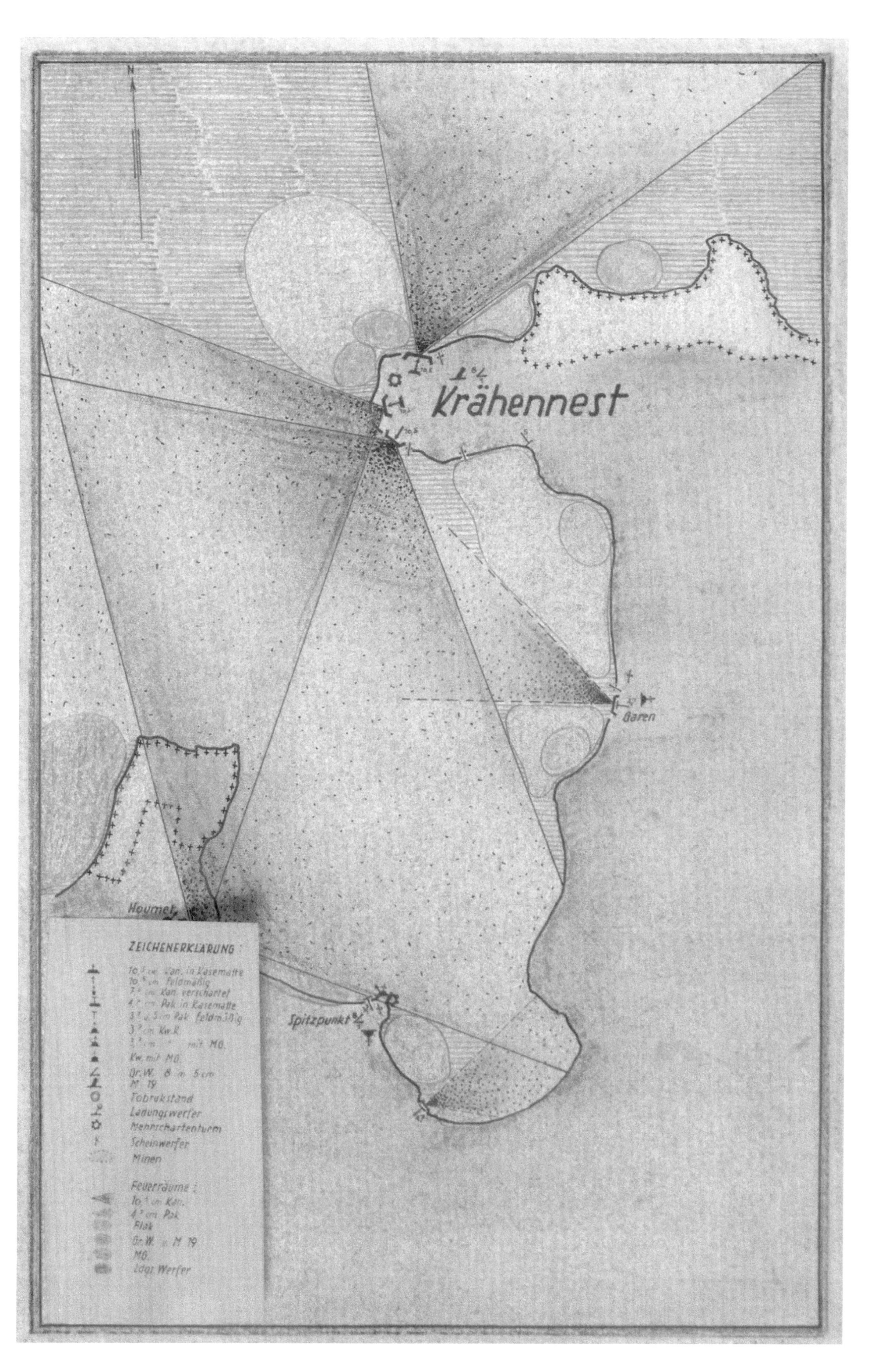
N
Krähennest
Bären
Houmet
Spitzpunkt
ZEICHENERKLÄRUNG:
Kan. in Kasematte
feldmäßig
Kan. verschartet
Pak in Kasematte
Pak feldmäßig
Kw.K.
mit MG.
Kw. mit MG.
Gr.W.
M 19
Tobrukstand
Ladungswerfer
Mehrschartenturm
Scheinwerfer
Minen
Feuerräume:
Kan.
Pak
Flak
Gr.W. u. M 19
MG.
Ldgs. Werfer

Le Grand Havre

Grossbucht

CHOUET Strong Point

1.) The strong point has been established in the sector 150 metres to the east of the direction-finding tower at Mont Cuet, up to road "Red 1" near infantry point 335.

2.) **Contingent**:

own forces:	1 officer	5 NCOs	40 men
assigned personnel for searchlight operation:		1 NCO	3 men

3.) **Weapons**:

own:
3 10.5 cm casemate guns (f), one in field position, two in fortified emplacements
1 multi-loopholed armoured turret (3 machine guns)
1 M19
1 8 cm mortar
1 3.7 cm anti-tank gun (d)
3 heavy machine guns
2 light machine guns
3 13.2 mm machine guns (b)
1 medium-range flame-thrower
10 defensive flame-throwers 42

assigned:
1 150 cm searchlight from Steinbruch Battery
2 60 cm searchlights (anti-aircraft artillery)

4.) **Military objectives**: The Chouet strong point is charged with preventing enemy landings in the assigned sector from sea or air. The combat sector of this strong point encompasses Le Grand Havre and the north and north-eastern parts of the strong point.

5.) **Operations**: The strong point will use all its weapons to repulse any attack from the sea. The main focus for the firepower is the entrance to Le Grand Havre and the possible landing places located there. Concentrations of fire from the weapons located at the strong point can be directed on these targets.

The entrance to the bay is covered by: two casemate guns, three heavy machine guns and two 13.2 mm machine guns (b). The beach is covered by: two casemate guns, one M19, one 8 cm mortar, two heavy machine guns and two 13.2 mm machine guns (b).

Small bays and areas not covered by flat trajectory fire can be covered by fire from the mortar.
The north-eastern part of the strong point is rocky and mined. This area is covered by: one casemate gun, one medium mortar, one M19, one 3.7 cm anti-tank gun, two machine guns and a 13.2 mm machine gun (b).

Regarding attacks from the land side, enemy approach is only possible from the Golf Course. Advances will be blocked by the strong point's reserve force manning the intermediate field positions. In the event of combat against superior enemy forces, this troop will retreat to the strong point, fighting as it does so. Airborne troops already landed will be repelled by sending an assault detachment.

Stützpunkt K R Ä H E N N E S T

1.) Der Stützpunkt ist eingesetzt im Abschnitt: 150 m ost=
wärts Peilturm Krähenberg bis Straße rot 1 bei I-Punkt 335

2.) Stärke:

eigene:	1 Offz.	5 Uffz.	40 Mannschaften
zugeteilte Scheinwerferbed.		1 Uffz.	3 Mannschaften

3.) Waffen:

eigene:	3	10,5 KK (f) davon 1 feldm., 2 festm.
	1	Mehrschartenturm (m.3 M.G.)
	1	M 19
	1	8 cm Gr.W.
	1	3,7 cm Pak (d)
	3	s,M.G.
	2	le.M.G.
	3	M.G. 13,2 mm (b)
	1	m.Fl.W.
	10	Abw.Flammenwerfer 42
zugeteilte:	1	150 cm Ø Scheinwerfer v.Steinbruch-Batterie
	2	60 cm Ø Scheinwerfer (Flak)

4.) Kampfauftrag: Der Stützpunkt Krähennest hat den Auftrag, jeden Landungsversuch von See und aus der Luft in dem zugewiesenen Abschnitt abzuwehren. Der Kampfabschnitt des Stützpunktes ist die Großbucht (Grand Havre), sowie der Nord- und Nordostteil des Stützpunktes.

5.) Kampfführung: Ein Angriff von See wird vom Stützpunkt aus unter Einsatz aller Waffen bekämpft. Schwerpunkt der Waffenwirkung liegt in der Einfahrt zur Großbucht und an deren Anlandemöglichkeiten. Dorthin können Feuerzu=
sammenfassungen der Stützpunktwaffen ausgelöst werden. In die Einfahrt wirken: 2 Kas.Kan., 3 s.M.G., 2 M.G.(b) 13,2 mm. Auf den Strand wirken: 2 Kas.Kan., 1 M 19, 1 8 cm Gran.W., 2 s.M.G., 2 M.G.(b) 13,2 mm.
Kleine Buchten und Stellen, die von Flachfeuer nicht be=
strichen werden, können durch Granatwerferfeuer ausge=
schaltet werden. Der Nordostteil des Stützpunktes ist felsenreich und außerdem vermint. In diesem Raum wirken: 1 Kas.Kan., 1 m. Gran.W., 1 M 19, 1 3,7 cm Pak, 2 M.G., 1 M.G.(b) 13,2 mm.
Bei Angriff von Land ist nur eine Annäherung des Feindes vom Golfplatz aus möglich. Abriegelung erfolgt durch Be=
setzung von Zwischenfeldstellungen durch eine Stützpunkt=
reserve. Bei Kampf gegen überlegenen Feind zieht sich diese Gruppe kämpfend auf den Stützpunkt zurück. Luftge=
landeter Gegner wird durch Entsenden eines Stoßtrupps be=
kämpft.

LA GARENNE Resistance Nest

1.) Deployment between the strong points at Chouet and Picquere Point.

2.) Contingent:

own:	1 NCO	9 men

3.) Weapons:

own:
1 4.7 cm anti-tank gun (t) coupled with machine gun (t)
1 machine gun (t)
1 machine gun MG311 (f) in tank turret
1 heavy machine gun MG34

4.) Military objectives: The La Garenne resistance nest is charged with preventing enemy landings in the assigned sector from the sea or air. The combat sector of the resistance nest encompasses Le Grand Havre and also, in the event of attack from the land side, the section of the Golf Course that is situated to the west of the road "Yellow 2".

5.) Operations: The resistance nest at La Garenne is sited on higher ground at the eastern shore of Le Grand Havre. The area at the entrance to Le Grand Havre will be covered by the 4.7 cm anti-tank gun (t). The beach to both sides of the resistance nest will be covered by the heavy machine gun and the machine gun (t). At the same time, these two weapons can be deployed – in conjunction with the machine gun MG311 in tank turret – to repulse an attack coming from the direction of the Golf Course. Should this happen, the intermediate field positions should be manned immediately by the resistance nest's assault detachment (consisting of one NCO and three men). In the event of combat against superior enemy forces, this detachment will retreat to the resistance nest, fighting as it goes. Airborne troops already landed will also be engaged by the assault detachment. Should a neighbouring strong point be attacked, it is to be supported by the assault detachment, if the situation permits.

Widerstandsnest G A R E N

1.) Einsatz zwischen den Stützpunkten Krähennest und Spitz=punkt.

2.) Stärke:
   eigene: 1 Uffz. 9 Mannschaften

3.) Waffen:
   eigene:
   1 4,7 cm Pak (t) m.gek.M.G.(t)
   1 M.G. (t)
   1 M.G. 311 (f) i.Pz.K.
   1 s.M.G.34

4.) Kampfauftrag: Das Widerstandsnest Garen hat den Auftrag, jeden Landungsversuch von See und aus der Luft in dem ihm zugewiesenen Abschnitt abzuwehren. Der Kampfabschnitt des Widerstandsnestes ist die Großbucht, sowie bei Angriff von der Landseite der westlich der Straße gelbe 2 gelege=ne Teil des Golfplatzes.

5.) Kampfführung: Das Widerstandsnest Garen liegt am erhöhten Strand der Ostseite der Großbucht. Die 4,7 cm Pak (t) wirkt in den Einfahrtsraum der Großbucht. Durch das s.M.G. und M.G. (t) wird der Strand beiderseits des Widerstands=nestes mit Feuer belegt. Gleichzeitig können diese beiden Waffen auch - zusammen mit dem M.G. 311 (f) in Panzer=kuppel - gegen einen Angriff vom Golfplatz her eingesetzt werden. In diesem Falle sofort Besetzung der Zwischen=feldstellungen durch die Stoßgruppe des Widerstandsnestes (Stärke 1/3). Bei überlegenem Feind zieht sich die Stoß=gruppe kämpfend auf das Widerstandsnest zurück. Luftge=landeter Gegner ist ebenfalls durch die Stoßgruppe zu bekämpfen. Wird ein Nachbarstützpunkt angegriffen, so ist dieser durch die Stoßgruppe zu unterstützen, sofern es die eigene Lage zulässt.

PICQUEREL POINT Resistance Nest

1.) Deployment between the resistance nest La Garenne and the resistance nest Le Houmet.

2.) Contingent:

own:	3 NCOs	18 men

3.) Weapons:

own:

1 multi-loopholed armoured turret (3 machine guns)
1 4.7 cm anti-tank gun (t) coupled with machine gun (t) 1
3.7 cm anti-tank gun
1 3.7 cm anti-tank gun (f) with machine gun MG311 (f) in tank turret
1 8 cm mortar
2 machine guns (one mounted on gun carriage 08)
1 machine gun 311 (f) in tank turret
3 defensive flame-throwers 42

4.) Military objectives: The Picquerel Point resistance nest is charged with preventing any enemy landing in the assigned sector from sea or air. Defence actions must focus on the southern part of Le Grand Havre.

5.) Operations: The resistance nest is situated in the southern part of Le Grand Havre and can use its weapons to cover the beach effectively to both sides of it. The 4.7 cm anti-tank gun (t) positioned in an emplacement below the former O.T. station will be used to repulse mechanised attacks.

The right-hand side is also covered by: one machine gun in the multi-loopholed armoured turret and one machine gun in a Tobruk pit. The left-hand side is covered by: one 3.7 cm anti-tank gun and two machine guns. Both sides are covered by: one 3.7 cm anti-tank gun in tank turret and one 8 cm mortar (especially to cover dead spaces). For rearward defence, the machine gun MG311 (f) in tank turret can be used. In the event of attacks from the land side, pre-prepared intermediate field positions will be manned immediately by an assault detachment from the resistance nest (consisting of one NCO and five men).

Widerstandsnest S P I T Z P U N K T

1.) Einsatz zwischen Widerstandsnest Garen und Widerstands=
nest Houmet.

2.) Stärke:
eigene: 3 Uffz. 18 Mannschaften

3.) Waffen:
eigene:
1 Mehrschartenturm (3 M.G.)
1 4,7 cm Pak (t) m.gek.M.G.(t)
1 3,7 cm Pak
1 3,7 cm Pak (f) m.M.G.311(f)i.Pz.Kuppel
1 8 cm Granatwerfer
2 M.G. (1 auf Lafette 08)
1 M.G.311 (f) i.Pz.Kuppel
3 Abwehrflammenwerfer 42

4.) Kampfauftrag: Das Widerstandsnest Spitzpunkt hat den Auftrag, jeden Landungsversuch von See und aus der Luft in dem ihm zugewiesenen Abschnitt abzuwehren. Schwerpunkt der Abwehr liegt im Südteil der Großbucht.

5.) Kampfführung: Das Widerstandsnest liegt im Südteil der Großbucht und hat nach beiden Seiten gute Waffenwirkung auf den Strand. Gegen Panzer wirkt die 4,7 cm Pak (t) aus einer Stellung unterhalb des früheren O.T.-Bahnhofs. Nach rechts wirken noch: 1 M.G. in Mehrschartenturm, 1 M.G. (Tobrukstand). Nach links wirken: 1 Pak 3,7 cm und 2 M.G. Nach beiden Seiten schießen: 1 Pak 3,7 cm in Pan= zerkuppel, 1 8cm Granatwerfer (besonders für tote Räume). Zur Rückwärtsverteidigung kann das M.G.311 (f) in Panzer= kuppel eingesetzt werden. Bei Angriff von Landseite wer= den vorbereitete Zwischenfeldstellungen durch eine Stoß= gruppe des Widerstandsnestes (Stärke 1/5) besetzt.

LE HOUMET Resistance Nest

1.) Deployment between the Picquerel Point and Pecqueries resistance nests.

2.) Contingent:

own:	3 NCOs	18 men
assigned (for searchlight operation):		2 men

3.) Weapons:

own:	1 10.5 cm casemate gun (f) in fortified emplacement
	1 5 cm anti-tank gun
	1 3.7 cm anti-tank gun (f) in tank turret
	2 light machine guns MG34
	2 machine guns MG311 (f) in tank turret
	1 8 cm mortar
	1 5 cm mortar
assigned:	1 60 cm anti-aircraft searchlight

4.) Military objectives: The Le Houmet resistance nest is charged with preventing any enemy landing in the assigned sector from sea or air.

The combat sector of the resistance nest encompasses the south-western part of Le Grand Havre, the Rousse Peninsula and, lying to the west of it, the Baie de Port Grat.

5.) Operations: The Le Houmet resistance nest is situated at the southern end of the Rousse peninsula, by the road "Red 2". It is bounded at the sides by two roads which lead to "Red 2". The casemate gun will cover Le Grand Havre and engage enemy forces landing at the beach. One machine gun MG311 (f) will cover the beach to the east-side of Rousse. The 5 cm and 3.7 cm anti-tank guns (in tank turrets) will be mainly deployed to fire at enemy targets (tanks) attempting to penetrate the terrain between the resistance nest at Le Houmet and the Steinbruch Battery, approaching from the small bay or from the road "Red 2" (from the west). The two machine guns will cover the same area. The 8 cm and 5 cm mortars will supplement the firepower of the other weapons and mainly target areas not covered by the flat-trajectory weapons.

The Steinbruch Battery will support the defensive fire in the area to the west of the resistance nest, using two 2 cm anti-aircraft machine guns, one 7.62 cm anti-tank gun (r) and one 3.7 cm anti-tank gun in tank turret. One machine gun MG311 (f) and one light machine gun MG34 will be deployed for rearward defence. In the event of attacks from the land side, pre-prepared intermediate field positions will be manned by an assault detachment from the resistance nest.

Widerstandsnest H O U M E T

1.) Einsatz zwischen den Widerstandsnestern Spitzpunkt und Fischerburg.

2.) Stärke:

eigene:	3 Uffz.	18 Mannschaften
zugeteilt(Schein=werferbedienung):		2 Mannschaften

3.) Waffen:

eigene:	1	1o,5 cm Kas.Kan.(f) festungsmäßig
	1	5 cm Pak
	1	3,7 cm Pak (f) in Panzerkuppel
	2	le.M.G.34
	2	M.G.311 (f) in Panzerkuppel
	1	8 cm Granatwerfer
	1	5 cm Granatwerfer
zugeteilt:	1	Flak-Scheinwerfer 60 cm.

4.) Kampfauftrag: Das Widerstandsnest Houmet hat den Auftrag, jeden Landungsversuch von See und aus der Luft in dem ihm zugewiesenen Abschnitt abzuwehren. Der Kampfabschnitt des Widerstandsnestes ist die Großbucht (Südwestteil), Halb=insel Rousse, sowie die westlich davon gelegene Baie de Port Grat.

5.) Kampfführung: Das Widerstandsnest Houmet liegt südlich der Halbinsel Rousse an der Straße Rot 2. Es ist seitlich von 2 Straßen, die auf die Rot 2 führen, begrenzt. Die Kas.Kan. wirkt in die Großbucht und bekämpft Anlandungen am Strand. 1 M.G. 311 (f) wirkt auf den Strand ostwärts Rousse. Die 5 cm und 3,7 cm Pak (in Panzerkuppel) haben den Auftrag, vor allem Feindziele (Panzer) zu bekämpfen, die aus der Klein-Bucht oder Straße Rot 2 (aus Westen) in das Gelände zwischen Widerstandsnest Houmet und der Batterie Stein=bruch einzubrechen versuchen. In den gleichen Raum wirken auch die beiden M.G. Der 8 cm und 5 cm Granatwerfer er=gänzen das Feuer der anderen Waffen und wirken vor allem in die von den Flachfeuerwaffen nicht bestrichenen Räume.

Die Steinbruch-Batterie unterstützt das Abwehrfeuer in den Raum westlich des Widerstandsnestes mit 2 2cm Fla.M.G., 1 7,62 cm Pak (r) und 1 3,7 cm Pak in Panzerkuppel. 1 M.G.311(f) und 1 le.M.G.34 sind zur Rückwärtsverteidi=gung eingesetzt. Bei einem Angriff von Land werden vorbe=reitete Zwischenfeldstellungen durch eine Stoßgruppe des Widerstandsnestes besetzt.

Top: Martello tower and old guardhouse at Chouet
Left Middle: One of the few remaining old guardhouses
Right Middle: Old fortification in a quarry

MARTELLO TURM UND ALTES WACHHAUS AUF KRÄHENNEST

EINES DER WENIGEN ALTEN WACHHÄUSER

IN EINEM STEINBRUCH EINE ALTE BEFESTIGUNGSANLAGE

Middle: Casemate gun at Chouet with and without camouflage

Bottom: View of Le Grand Havre from the casemate gun at Mont Cuet. To the left, the Martello tower near the Le Houmet strong point

K.K. AUF KRÄHENNEST MIT UND OHNE TARNUNG

BLICK VON DER K.K. KRÄHENBERG AUF DIE GROSSBUCHT · LINKS DES MARTELLO TURM STÜTZP. HOUMET

Middle: Loopholed turret at Chouet, well camouflaged

Bottom: Inside the loopholed turret: view from the control post to the magazine

GUT GETARNTER SCHARTENTURM AUF KRÄHENNEST

IM SCHARTENTURM · BLICK VOM KAMPFSTAND IN DEN MUNITIONSRAUM

Middle: Camouflaged bunkers at Chouet

Bottom: La Bennette near Chouet

GETARNTE BUNKER AUF KRÄHENNEST

LA BENNETTE BEI KRÄHENNEST

Top: Le Grand Havre from Chouet to Vale Church
Middle: La Garenne at Le Grand Havre
Bottom: Le Grand Havre with the Chouet strong point and La Garenne

DIE GROSSBUCHT VON KRÄHENNEST ZUR VALE KIRCHE

ORTSTEIL LA GARENNE AN DER GROSSBUCHT

DIE GROSSBUCHT MIT STÜTZP. KRÄHENNEST UND GAREN

Top: At Picquerel Point, a multi-loopholed armoured turret camouflaged as a rock
Middle: Tank turret with 3.7 cm anti-tank gun (f) and machine gun MG311 (f) at Picquerel Point
Bottom: Bunker and multi-loopholed armoured turret at Picquerel Point

ALS FELSEN GETARNTER MEHRSCHARTER AUF SPITZPUNKT

PANZERKUPPEL MIT 3,7 PAK (F) U. MG 311 (F) AUF SPITZPUNKT

BUNKER UND MEHRSCHARTER AUF SPITZPUNKT

Top: View of the Le Houmet resistance nest from Picquerel Point
Middle: Emplacement of the casemate gun at Le Houmet, camouflaged as a cottage
Bottom: View from Le Houmet to Steinbruch Battery. In the foreground, 3.7 cm anti-tank gun (f) in tank turret

WIEDERSTANDSNES HOUMET VON SPITZPUNKT AUS

BUNKER DER K.K. AUF HOUMET
ALS LANDHAUS GETARNT

BLICK VON HOUMET ZUR
BATTERIE STEINBRUCH
VORN 3,7 PAK(F) IN PZ.-KUPPEL

View across the north-western part of the island. To the left Vale Church, Chouet and Fort Le Marchant

To subscribe and support this part work edition please
visit our website where you will find details of how
to subscribe and how the edition will unfold.

www.clearvuepublishing.com

or email us at: info@clearvuepublishing.com

Alternatively you can write to us:
The Clear Vue Publishing Partnership Ltd,
La Battue, St Peter Port,
Guernsey,
GY1 1UP

Our film on the history of the fortifications,
Hitler's Island Madness,
features images from the Festung
and is available in shops
and on our website
as is our three part oral history
The Occupation of the Channel Islands

Visitors to the Channel Islands can
Find out more about the fortifications at:

The Channel Islands Occupation Society [Guernsey]
www.occupied.guernsey.net

The Channel Islands Occupation Society [Jersey]
www.ciosjersey.org.net

Festung Guernsey
www.festungguernsey.supanet.com

First published 2007
in a limited edition of 135 copies

This paperback partwork edition
Published 2012

The Clear Vue Publishing Partnership Ltd
La Battue, Candie Road, St Peter Port
Guernsey, Channel Islands
GY1 1UP

Festung Guernsey reproduced by kind permission of: The Royal Court 2007

www.clearvuepublishing.com